NUREG-1700, Rev. 1

Standard Review Plan for Evaluating Nuclear Power Reactor License Termination Plans

I0467982

Manuscript Completed: March 2003
Date Published: April 2003

Prepared by
C. L. Pittiglio

Division of Waste Management
Office of Nuclear Material Safety and Safeguards
U.S. Nuclear Regulatory Commission
Washington, DC 20555-0001

Abstract

This standard review plan (SRP) guides the staff of the U.S. Nuclear Regulatory Commission (NRC) in conducting safety reviews of license termination plans (LTPs). The principal purpose of this SRP is to ensure the quality and uniformity of NRC staff reviews and to present a well-defined base from which to evaluate the requirements for terminating the license of a nuclear power plant. It is also the purpose of this SRP to make information about regulatory matters widely available, so that interested members of the public and the nuclear industry can gain a better understanding of the staff's review process. Licensees may also use this SRP to assist in developing an LTP. Each SRP presents the acceptance criteria for all areas of review for license termination, and identifies the matters to be reviewed, the basis for the review, and the conclusions that are sought. For example, NUREG-1757, "Consolidated NMSS Decommissioning Guidance" provides additional guidance on how to demonstrate compliance with the unrestricted release, restricted release, and alternative criteria for license termination, in accordance with Subpart E of Title 10, Part 20, of the Code of Federal Regulations (10 CFR Part 20). When the three volumes of NUREG-1757 are complete, the consolidated guidance will completely replace NUREG-1727. To avoid duplication of information, this SRP references applicable sections of NUREG-1757 for additional guidance. Once NUREG-1757 is complete, the NRC will revise this SRP to reference the additional sections of NUREG-1757, if necessary. In addition, the NRC will also issue supplemental guidance which will subsequently be consolidated into NUREG-1757 to address partial site release prior to LTP approval, and compliance with Subpart E.

Paperwork Reduction Act Statement

The information collections contained in this NUREG are covered by the requirements of 10 CFR Parts 20, 30, 40, 50, 51, 70, and 72 which were approved by the Office of management and Budget, approval numbers 3150-0014, 0017, 0020, 0011, 0021, 0009 and 0132

Public Protection Notification

NOTE: This document was completely rewritten. Do not use NUREG- 1700. Only use NUREG-1700, Revision 1.

Contents

Abbreviations

ALARA	as low as is reasonably achievable
CFR	*Code of Federal Regulations*
DCGL	derived concentration guidelines
DG	draft regulatory guide
EIS	environmental impact statement
ER	environmental report
FR	*Federal Register*
FSAR	final safety analysis report
ISFSI	independent spent fuel storage installation
LTP	license termination plan
NMSS	Office of Nuclear Material Safety and Safeguards (NRC)
NRC	U.S. Nuclear Regulatory Commission
PSDAR	post–shutdown decommissioning activities report
QA	quality assurance
RG	regulatory guide
SE	safety evaluation
SRP	standard review plan

A. Introduction

1. PURPOSE

This standard review plan (SRP) is intended to guide the staff of the U.S. Nuclear Regulatory Commission (NRC) in conducting safety reviews of license termination plans (LTPs) for nuclear power reactors. The principal purpose of this SRP is to ensure the quality and uniformity of NRC staff reviews and to present a well-defined base from which to evaluate the requirements for terminating the license of a nuclear power plant. It is also the purpose of this SRP to make information about regulatory matters widely available, so that interested members of the public and the nuclear industry can gain a better understanding of the staff's review process. Licensees may also use this SRP to assist in developing an LTP. SRPs are not substitutes for regulations, and compliance with them is not required. Methods and solutions different from those set out in this SRP will be acceptable if they provide a basis for concluding that the LTP is in compliance with the Commission's regulations.

In general, the NRC staff bases its license termination review primarily on the information that the applicant submits. Consequently, the LTP should contain sufficient detail to enable the staff to independently verify that the facility can be decommissioned safely, and the license can be terminated. The specific information that the staff needs in order to evaluate an LTP is identified in Regulatory Guide (RG) 1.179, "Standard Format and Content of License Termination Plans for Nuclear Power Reactors" (Ref. 1). Because the LTP must be submitted two years or more prior to license termination, the level of detail required in the LTP will vary depending on when the LTP is submitted.

In order to implement NRC's streamlined approach for licensing actions, when a licensee submits an LTP to the NRC for review and approval, the staff will use this SRP to evaluate the information submitted by licensees to support the evaluation of the decommissioning of its facility. The staff's evaluation will include: (1) Acceptance Reviews, (2) Detailed Safety Reviews, (3) Requests for Additional Information (RAIs) if needed, and (4)Environmental Report Reviews. The staff shall ensure the application is complete by conducting an acceptance review, and if it is not, return it to the licensee. Appendix 1, "Acceptance Review Checklist for Unrestricted or Restricted Release of the Site" provides a checklist of information based on RG-1.179 that needs to be addressed in the LTP, and is used by the staff to determine if the licensee has included all the information required in the LTP. The information identified in the checklist will vary depending upon the amount of decommissioning that has been completed when the LTP is submitted. If the application is complete, staff will then conduct a detailed review, and prepare its preliminary technical evaluation. Through this process, staff will be able to identify areas where issues need to be addressed. This approach will help ensure that questions are limited to those areas where additional information is needed, and should help reduce the number of questions.

2. REGULATIONS AND RELATED GUIDANCE

On July 29, 1996, the Commission published a *Federal Register* notice regarding amendments to its regulations in 10 CFR Parts 2, 50, and 51of Title 10 of the Code of Federal Regulations (10 CFR) (61 FR 39278) (Ref. 2). In general, the purpose of those amendments was to prescribe specific criteria for decommissioning nuclear power reactors, which became effective on August 28, 1996. This rule, by eliminating, revising, or extending operating reactor requirements to be commensurate with the necessary level of safety, specifies requirements for reactors that present a significantly reduced risk to the public because they are permanently shut down and no longer have fuel in the reactor vessel.

Decommissioning activities for power reactors may be divided into three phases including: (1) initial activities, (2) major decommissioning and storage activities, and (3) license termination activities. Regulatory Guide 1.184, "Decommissioning of Nuclear Power Reactors" (Ref. 3) describes acceptable methods and procedures for implementing the rules that relate to Phases 1 and 2.

For Phase 3, 10 CFR 50.82(a)(9) specifies that an application for license termination must be accompanied or preceded by an LTP, which is subject to NRC review and approval. According to 10 CFR 50.82(a)(9), the licensee must submit an LTP at least 2 years before termination of the license, the NRC approves the LTP by issuing a license amendment, and NRC must hold a public meeting near the site. Any hearing held in relation to an LTP would fall under either Subpart G or Subpart L of 10 CFR Part 2. If an applicant submits an LTP while the spent fuel is stored under the Part 50 license, Subpart G of 10 CFR Part 2 would apply. Conversely, if the applicant has permanently moved the fuel to an authorized facility, a hearing on the proposed the LTP would be in accordance with Subpart L.

In accordance with 10 CFR 50.82(a)(10), the LTP is approved by license amendment. Recognizing that there may be a need to make changes to the LTP during decommissioning, the licensee should include a provision in the LTP that addresses changes to the LTP after approval. Appendix 2, "LTP Areas that Cannot Be Changed Without NRC Approval" sets out such a provision that the NRC finds acceptable.

On July 21, 1997, the Commission amended its regulations in 10 CFR Parts 20, 30, 40, 50, 51, 70, and 72 (62 FR 39058) (Ref. 4), prescribing specific radiological criteria for license termination. The license termination rule (10 CFR Part 20, Subpart E) requires the licensee to evaluate the entire site for compliance with Subpart E at the time of license termination. To ensure that the entire site meets the radiological release requirements of 10 CFR Part 20, Subpart E, at the time the license is terminated, and to avoid licensees taking a piecemeal approach to license termination, the LTP should consider the entire site as defined in the original license or FSAR, along with any subsequent additions to the site boundary. In addition, licensees should be aware of the record keeping requirements in the soon to be promulgated 10 CFR 50.83, "Release of Part of a Facility or Site for Unrestricted Use" (Ref. 5), (66 FR 46230), and other applicable areas when planning a partial release prior to license termination to ensure that the dose contribution will not adversely impact future partial release plans or the final release at license termination.

To avoid duplication of information, this SRP references applicable sections of NUREG-1757, "Consolidated NMSS Decommissioning Guidance" (Ref. 6), which provides additional guidance on how to demonstrate compliance with the unrestricted release, restricted release, and alternative criteria for license termination in accordance with Subpart E of 10 CFR Part 20. When the three volumes of NUREG-1757 are complete, the consolidated guidance will replace NUREG-1727, the NMSS Decommissioning SRP. In addition, the NRC will issue supplemental guidance which will subsequently be consolidated into NUREG-1757, to address partial site release prior to LTP approval, and compliance with Subpart E dose constraints.

3. LTP APPLICABLE REQUIREMENTS

As discussed in the statements of consideration (SOCs) (61 FR 39278) (Ref. 2) that accompanied the Final Rule on Decommissioning of Nuclear Power Reactors, the Commission must make decisions regarding the licensee-proposed actions described in the LTP. In particular, the Commission must evaluate (1) the licensee's plan for ensuring that sufficient funds will be available (2) proposed residual radioactivity levels for license termination, and (3) the adequacy of the final survey to verify residual radioactivity levels have been met. To support the Commission's findings, the LTP must contain a dose assessment, a final survey plan, and a decommissioning cost estimate. Other information in the LTP supports the staff's evaluation of these areas. Because the LTP must be submitted two years or more prior to license termination, the level of detail required to be submitted in the LTP will vary depending on when the licensee submits the LTP. The information submitted in the LTP should reflect the current status of the decommissioning at the facility. 10 CFR 50.82(a)(9)(ii) requires that the LTP must include the following information:

- site characterization

- identification of the remaining dismantlement activities

- plans for site remediation

- detailed plans for the final radiation survey

- description of the end use of the site, if restricted

- an updated site-specific estimate of remaining decommissioning costs

- a supplement to the environmental report describing any new information or significant environmental change associated with the licensee's proposed termination activities

The LTP must be submitted as a supplement to the licensee's final safety analysis report (FSAR) or as an equivalent document. A licensee might submit the LTP concurrently with the post-shutdown decommissioning activities report (PSDAR). Guidance on the content of the PSDAR can be found in Regulatory Guide 1.185, "Standard Format and Content Guide for Post-Shutdown Decommissioning Activities Report" (Ref. 7).

Currently, the Division of Waste Management, Decommissioning Branch of the Office of Nuclear Material Safety and Safeguards (NMSS) is responsible for reviewing the LTP, and developing this SRP.

4. ORGANIZATION OF THIS SRP

The SRP is written to cover a variety of license termination conditions. Each section of the SRP presents the acceptance criteria for all areas of review for license termination. It identifies the matters to be reviewed, the basis for the review, and the conclusions that are sought. For a given application, the staff reviewers will select appropriate aspects of each section based on the status of the decommissioning. In some cases, a facility feature may be sufficiently similar to a previously reviewed feature so that a complete new review is not needed.

The remainder of this SRP is divided into the three sections including: (B) LTP Standard Review Plan and Acceptance Criteria (C) Evaluation Findings; and (D) References.

(B)   "LTP Standard Review Plan and Acceptance Criteria

Each subsection in Section B, "LTP Standard Review Plan and Acceptance Criteria," summarizes the purpose of the review and the applicable NRC requirements. The initial summary is not designated "Area of Review," as described in detail in RG 1.179 (Ref. 1) because the Area of Review and Acceptance Criteria sections were combined into one section. Each subsection includes applicable acceptance criteria, which contain the technical bases for assessing the acceptability of the analysis or program. The technical bases include references such as NRC regulatory guides, industry codes and standards, and branch technical positions. These approaches are listed in the SRP so the staff can take consistent positions on similar problems as they arise. Branch technical positions and Regulatory Guides present approaches that are acceptable to the staff for demonstrating compliance with the NRC regulatory requirements, but are not considered to be the only possible approaches. However, licensees proposing approaches other than those described in the branch technical positions or Regulatory Guides may expect longer review times and the staff may need additional justification/analysis before approving the proposed approach.

(C)   Evaluation Findings

This section presents the type of conclusion that is sought by the staff for the particular review area. For each area of review, the safety evaluation (SE) will include a description of the review, including aspects of the review that were selected or emphasized, and the bases for any deviation from the SRP.

(D)   References

The section presents the references used as part of the supporting basis for the technical conclusion reached.

B. LTP Standard Review Plan and Acceptance Criteria

1. General Information

The LTP must address each of the areas delineated in 10 CFR 50.82(a)(9) as well as Subpart E of 10 CFR Part 20. The regulations applicable to this review are 10 CFR 50.82(a)(9), 10 CFR 50.82(a)(10), and Subpart E of 10 CFR Part 20.

Acceptance Criteria

- The LTP is submitted in the form of a supplement to the FSAR or equivalent and the LTP has preceded or is accompanied by an application for license termination.

- The LTP is submitted 2 years or more before the proposed termination date of the license.

- The LTP is submitted in the form of a license amendment request.

- The LTP lists the name and address of the licensee; license number; docket number; facility name and address; size of the site in acres or square meters; the State and county in which the site is located; the names of and distances to nearby communities, towns, and cities; a description of the contours and features of the site; the elevation of the site; a description of property surrounding the site, including the location of all off-site wells used by nearby communities or individuals; the location of the site relative to prominent features such as rivers and lakes; a map that shows the detailed topography of the site using a contour interval; the location of the nearest residences and all significant facilities or activities near the site; and a description of the facilities (buildings, parking lots, fixed equipment, etc.) at the site.

- The LTP identifies all changes to the site boundaries (as defined in 20.1003) that have occurred. Forthcoming changes to 10 CFR 50.75(g) as a result of the partial site release rulemaking will require licensee's to keep records that document any changes to the original site boundary.

- The LTP addresses each of the following 10 CFR 50.82(a)(9) areas as well as the radiological criteria for unrestricted or restricted release of the site as defined in Subpart E of 10 CFR Part 20:

  - site characterization

  - identification of remaining site dismantlement activities

  - plans for site remediation

  - detailed plans for final radiation surveys for release of the site

  - method for demonstrating compliance with the radiological criteria for license termination. For restricted release, the LTP should also describe the site's end use, and provide documentation demonstrating compliance with the public consultation, institutional controls, and financial assurance requirements of 10 CFR 20.1403 or 10 CFR 20.1404.

  - updated site–specific estimate of remaining decommissioning costs

  - supplement to the environmental report, pursuant to 10 CFR 51.53(d), which describes any new information or significant environmental changes associated with the licensee's proposed termination activities

2. Site Characterization

Site characterization information is provided to determine the extent and range of radioactive contamination on site, including structures (on a structure by structure basis and as necessary on a room by room basis), systems, components, residues, soils, and surface and ground water. On the basis of the site characterization, the licensee designs final radiation surveys to evaluate all areas in which contamination previously existed, remains, or has the potential to remain.

The licensee should also use the site characterization information to develop input for use in the dose modeling. As part of the review, the NRC staff should review the licensee's site characterization plans and site records (required under 10 CFR 50.75(g)). The purpose of this review is twofold. First, the staff seeks to ensure that the site characterization presented in the LTP is complete. Second, the staff verifies that the licensee obtained the data using sufficiently sensitive instruments and proper quality assurance procedures to obtain reliable data that are relevant to determining whether the site will meet the decommissioning limits if characterization data is used as final survey data. The regulation applicable to this area of review is 10 CFR 50.82(a)(9)(ii)(A).

Additional guidance on site characterization and characterization surveys can be found in Section 4 of NUREG-1757, Volume 2 (Ref. 6). In addition, if the licensee proposes site-specific derived concentration guidelines (DCGLs), the hydrogeologic information described in NUREG-1757 will likely be required to support the parameters used in the site-specific dose assessment.

Acceptance Criteria

- The LTP identifies all locations, both inside and outside the facility, where radiological spills, disposals, operational activities, or other radiological accidents and or incidents occurred and could have resulted in contamination. This should be done on a room by room or area by area basis as necessary, including equipment, laydown areas, or soils (subfloor and outside area).

- The LTP describes, in summary form, the original shutdown and current radiological and non-radiological status of the site.

- The LTP site characterization is sufficiently detailed to allow the NRC staff to determine the extent and range of radiological contamination of structures, systems (including sewer systems and waste management systems), floor drains, ventilation ducts, piping and embedded piping, rubble, ground water and surface water, components, residues, and environment, including maximum and average contamination levels and ambient exposure rate measurements of all relevant areas (structures, equipment, and soils) of the site (including contamination on and beneath paved parking lots).

- The LTP identifies the survey instruments and supporting quality assurance practices used in the site characterization program.

- The LTP identifies the background levels used during scoping or characterization surveys.

- The LTP describes in detail the areas and equipment that need further remediation to allow the reviewer to estimate the radiological conditions that will be encountered during remediation of equipment, components, structures, and outdoor areas.

3. Identification of Remaining Site Dismantlement Activities

The LTP describes the remaining site dismantlement activities. The regulation applicable to this area of review is 10 CFR 50.82(a)(9)(ii)(B).

Acceptance Criteria

- The LTP discusses the remaining tasks associated with decontamination and dismantlement, estimates the quantity of radioactive material to be shipped for disposal or processing, describes the proposed control mechanisms to ensure that areas are not recontaminated, and contains occupational exposure estimates and radioactive waste characterization.

- The LTP describes the remaining dismantlement activities in sufficient detail for the NRC staff to identify any associated inspection or technical resources that will be needed.

- The LTP is sufficiently detailed to provide data for use in planning further decommissioning activities. As such, the LTP includes decontamination techniques, projected schedules, costs, waste volumes, dose assessments (including groundwater assessments), and health and safety considerations.

- The LTP lists the remaining activities that do not require any additional licensing action.

4. Remediation Plans

The LTP discusses in detail how facility and site areas will be remediated to meet the NRC's release criteria. The regulations applicable to this area of review are 10 CFR 50.82(a)(9)(ii)(C) and Subpart E of 10 CFR Part 20 (Ref. 3).

Additional guidance on this topic can be found in Section 4 of NUREG-1757, Vol. 2 (Ref. 6).

Acceptance Criteria

- The LTP addresses any changes in the radiological controls to be implemented to control radiological contamination associated with the remaining decommissioning and remediation activities.

- The LTP discusses in detail how facility and site areas will be remediated to meet the proposed residual radioactivity levels (DCGLs) for license termination. Discussions should focus on any unique techniques or procedures used to evaluate whether the DCGLs have been met including the following:

    - Summarize the techniques that will be used to remediate building structures and components (e.g., scabbling, hydrolazing, grit blasting, etc.).

    - Summarize the equipment that will be decontaminated and how the decontamination will be accomplished.

    - Summarize the radiation protection methods and control procedures that will be employed including a summary of the procedures already authorized under the existing license.

    - Commit to conduct decommissioning activities in accordance with approved written procedures.

    - Include a detailed description of the techniques that will be employed to remove or remediate surface and subsurface soils, groundwater, and surface water and sediments.

    - Describe plans, if any, for onsite disposal of decommissioning waste.

- The LTP includes a schedule that demonstrates how and in what time frames the licensee will complete the interrelated decommissioning activities. 10 CFR 50.82 (a)3 requires completion of decommissioning within 60 years. If the completion of decommissioning is delayed for more than 60 years, the LTP, includes a justification for the delay in accordance with 10 CFR 50.82(a)(3).

5. Final Radiation Survey Plan

The LTP describes the final radiation survey plan for demonstrating that the plant and site will meet the proposed release limits. The regulations applicable to this area of review are 10 CFR 50.82(a)(9)(ii)(D) and 10 CFR 20.1501(a) and (b). The final status survey is the radiation survey performed after an area has been fully characterized, remediation has been completed, and the area is ready to be released. The purpose of the final status survey is to demonstrate that the area conforms to the radiological criteria for license termination. The final status survey is not conducted for the purpose of locating residual radioactivity; the historical site assessment and the characterization survey perform that function.

Additional guidance for final survey plans is contained Section 4, NUREG-1757, Vol. 2 (Ref. 6)

Acceptance Criteria

- The LTP includes the "Information To Be Submitted," as described in section 4 of NUREG-1757 Vol 2. In addition the LTP includes the following information:

    - identification of the major radiological contaminants

    - methods used for addressing hard-to-detect radionuclides

    - access control procedures to control recontamination of clean areas

    - description of the Quality Assurance (QA) Program, to support both field survey work and laboratory analysis, which addresses the QA organization; training and qualification requirements; survey instructions and procedures, including water, air, and soil sampling procedures; document control; control of purchased items; inspections; control of survey equipment; handling, storage, calibration, and response checks; shipping of survey equipment and laboratory samples; nonconformance items; corrective action; QA records; and the survey audits, including methods to be used for reviewing, analyzing, and auditing data

    - methods for surveying embedded piping

- The final survey plan meets the evaluation criteria defined in Section 4 of NUREG-1757, Vol. 2 (Ref. 6).

6. Compliance With the Radiological Criteria for License Termination

The licensee must clearly present in the LTP the radiological criteria proposed for license termination. If a licensee desires an unrestricted release in accordance with the requirements of 10 CFR 20.1402, the LTP should describe the methods used to demonstrate compliance.

If a licensee requests license termination under the restricted release criteria in accordance with Subpart E of 10 CFR Part 20, the LTP should describe in detail how the requirements of 10 CFR 20.1403 and 10 CFR 50.82(a)(9)(ii)(E) will be met. If a licensee requests license termination under the alternative radiological criteria in 10 CFR 20.1404, the LTP should describe how the requirements in 10 CFR 20.1301(a)(1), 10 CFR 20.1404, and 10 CFR 50.82(a)(9)(ii)(E) will be met.

The information that should be submitted in the LTP and the associated evaluation criteria are described in NUREG-1727 (Ref. 6). The following table identifies the applicable portions of this NUREG.

Application	Appropriate Sections of NUREG-1757, Vol. 2
Unrestricted release using screening criteria	5.1 and Appendix H
Unrestricted release using site-specific information	5.2 and Appendix I
Restricted release	5.3, and Appendices I, and J
Alternative criteria	5.4, and Appendices I, and J
As Low As is Reasonably Achievable (ALARA)	6.0 and Appendix N

7. Update of the Site-Specific Decommissioning Costs

The LTP must provide an estimate of the remaining decommissioning costs for unrestricted or restricted release of the site. The LTP must also compare the estimated costs with the present funds set aside for decommissioning, and must note how the financial assurance instruments, which are required for decommissioning under 10 CFR 50.75, will be increased, if necessary. The financial assurance instruments required under 10 CFR 50.75 must be funded to the amount of the cost estimate. If there is a deficit in current funding, the LTP must indicate the means for ensuring adequate funds to complete decommissioning. The regulation applicable to this area is 10 CFR 50.82(a)(9)(ii)(F).

The NRC does not consider decommissioning costs to include the costs for demolition of decontaminated structures, site restoration activities, or other activities that are not involved with removing the facility from service or reducing residual radioactivity. Rather, the NRC considers such costs to be utility operating expenses and as such, these expenses are not included in the amount of money required to be placed in the plant's decommissioning fund in accordance with 10 CFR 50.75. The costs of constructing, operating, maintaining, and decommissioning an onsite spent fuel storage facility or ISFSI are also explicitly excluded from decommissioning costs. A licensee is, however, required to separately notify the NRC of its program to manage and provide funding for the management of irradiated fuel.

The NRC has revised this section of NUREG-1700 to make it consistent with NUREG-1713, "Standard Review Plan for Decommissioning Cost Estimates for Nuclear Power Reactors" (Ref. 8) and Draft Regulatory Guide (DG)-1085, "Standard Format and Content of Decommissioning Cost Estimates for Nuclear Power Reactors" (Ref. 9) as they are related to developing a detailed site-specific cost estimate required by 10 CFR 50.82(a)(8)(iii). The licensee will only be required to update that site-specific cost estimate to reflect any changes that occurred since it was initially submitted. For example, the licensee could be required to update the LTP cost estimate to reflect completed decommissioning activities, inflation, and changes in radioactive waste disposal cost. If little

14

decommissioning has been completed, and inflation and disposal costs have not changed, the cost estimate required by 10 CFR 50.82(a)(8)(iii) may be acceptable. The NRC is not requiring the licensee to submit contractual documents or agreements that exist between the licensee and its decommissioning contractor, and the cost estimate should not be effected by the licensee's election to decommission the facility itself or to contract the decommissioning of the facility to another party. When the licensee is required to submit an update to the site-specific cost estimate required by 10 CFR 50.82(a)(8)(iii), the update should reflect the current status of the facility, and the licensee's plans for how the actions will be completed. Because the financial assurance instrument required under 10 CFR 50.75 must be funded to the amount of the cost estimate, and because the licensee has been allowed to withdraw the allocated funds during decommissioning, the updated site-specific cost estimate must address the remaining activities necessary to complete decommissioning in order to ensure that sufficient funds are available.

Acceptance Criteria

- The LTP decommissioning cost estimate includes an evaluation of the following cost elements:

Cost Elements

- cost assumptions used, including a contingency factor (normally 25%)

- major decommissioning activities and tasks

- unit cost factors

- estimated costs of decontamination and removal of equipment and structures

- estimated costs of waste disposal, including applicable disposal site surcharges and transportation costs

- estimated final survey costs

- estimated total costs

15

- The LTP focuses on detailed activity by activity cost estimates.

- The LTP also compares the funds available for decommissioning with the calculated total cost from the licensee's detailed cost analysis. In addition, Regulatory Guide 1.159, "Assuring the Availability of Funds for Decommissioning Nuclear Reactors" (Ref. 10), explains in detail the methods for estimating decommissioning costs, as well as accepted financial assurance mechanisms.

- The LTP cost estimate is based on credible engineering assumptions, and the assumptions are related to all major remaining decommissioning activities and tasks and are consistent with the information identified in Sections B3 and B4 of this SRP.

- The LTP cost estimate includes the cost of the remediation action being evaluated, the cost of transportation and disposal of the waste generated by the action, and other costs that are appropriate for the specific case. The current version of NUREG-1307, "Report on Waste Burial Charges" (Ref. 11), provides guidance on estimating waste disposal costs.

8. Supplement to the Environmental Report

The licensee must submit a supplement to the environmental report (ER) describing any new information or significant environmental changes associated with the site-specific termination activities. The licensee should not limit the focus of the environmental review to radiological activities related to decommissioning activities, and should describe in detail the proposed termination activities from the time the LTP is submitted until the license is terminated. The review should focus, but not be limited to, activities associated with the radiological decontamination of the facility. The supplement to the ER should include a detailed description of remaining activities, the interaction between those activities and the environment, and the likely environmental impact of those activities. The supplement to the ER should also state the licensee's determination regarding whether the activities and their impacts are bounded by the impacts predicted by their own site specific EIS developed in support of licensing the facility, NUREG-0586 as supplemented (Ref. 12), or the PSDAR. In addition, the supplement to the ER should include a determination on any site-specific environmental assessments, as well as a description of the approach the licensee will use if the impacts exceed the impacts as a result of licensing actions of existing environmental assessments. The licensee's determination should be supported by a detailed comparison of the proposed impacts to predictions of the impacts in the existing environmental assessments. The regulations applicable to this area of review are 10 CFR 50.82(a)(9)(ii)(G) and 10 CFR 51.53. For restricted release, the supplement to the ER will be addressed on a case-by-case basis.

Acceptance Criteria

- The supplement to the ER describes changes to the data that have arisen since the licensee issued its updated "Environmental Report—Operating License Stage" related to the site location, climate, demography, socioeconomic data, land use, surface water, ground water, and biota.

- The supplement to the ER describes the impacts associated with site-specific termination activities from the time the LTP is submitted until the license is terminated.

- The supplement to the ER states the licensee's determination regarding whether the activities and their impacts are bounded by the impacts predicted by their own site-specific EIS developed in support of licensing the facility, NUREG-0586 as supplemented, the PSDAR, or any existing site-specific environmental assessments. The Supplement to the ER also describes the approach that the licensee will use if the impacts exceed the impacts in existing environmental assessments.

C. Evaluation Findings

In reviewing an LTP, the NRC staff will consider whether the licensee has met each of the requirements set out below and whether the plan provides an adequate basis for each of the following findings identified below and these findings are founded on the Acceptance Criteria defined in Sections B. 1- 8 of this SRP.

The Evaluation Findings are as follows:

- The licensee submitted the LTP as a supplement to the facility's FSAR or its equivalent in accordance with 10 CFR 50.82(a)(9)(i).

- The licensee met the objective of providing an adequate site characterization as required by 10 CFR 50.82(a)(9)(ii)(A).

- The licensee identified the remaining site dismantlement activities that are necessary to complete the decommissioning of the facility, as required by 10 CFR 50.82(a)(9)(ii)(B).

- The licensee adequately described its plans for site remediation, as required by 10 CFR 50.82(a)(9)(ii)(C).

- The licensee's final radiation survey plan adequately demonstrates that the plant and site will meet the radiological release criteria for license termination as defined in 10 CFR 50.82(a)(9)(ii)(D).

- The licensee adequately described how it will meet the requirements of 10 CFR 50.82(a)(9)(ii)(E), with respect to the end-use of the site, if the licensee requests license termination under the restricted release criteria.

- The licensee met the requirements of 10 CFR 50.82(a)(9)(ii)(F) by providing an updated site-specific estimate of the remaining decommissioning costs, and plans for ensuring the availability of adequate funds for decommissioning.

- The licensee met the requirements of 10CFR 50.82(a)(9)(ii)(G) and 10 CFR 51.53 by providing acceptable updates to the "Environmental Report Operating License Stage."

- For unrestricted release, the licensee met the requirements of 10 CFR 20.1402, in that the LTP demonstrates the radiological criteria for unrestricted release will be met.

- For restricted release, the licensee met the requirements of 10 CFR 20.1403(a), in that the LTP demonstrates further reductions in residual radioactivity to allow the site to be released the site for unrestricted use either (a) would result in net public or environmental harm, or (b) are not being made because the residual levels are ALARA.

- For restricted release, the licensee met the requirements of 10 CFR 20.1403(b), in that the LTP demonstrates that with the institutional controls in place, the dose criteria for restricted release will be met.

- For restricted release, the licensee met the requirements of 10 CFR 20.1403(c), in that the LTP demonstrates that sufficient financial assurance is available to enable a third party to assume and carry out any necessary maintenance of the site.

- For restricted release, the licensee met the requirements of 10 CFR 20.1403(d), in that the LTP demonstrates the requirements for public involvement have been met.

- For restricted release, the licensee met the requirements of 10 CFR 20.1403(e), in that the LTP demonstrates that the dose criteria for restricted release will still be met in the event that the instructional controls should fail.

- For license termination using alternative criteria, the licensee met the requirements of 10 CFR 20.1404(a), in that the LTP demonstrates the dose criteria for license termination using alternative criteria will be met.

- For license termination using alternative criteria, the licensee met the requirements of 10 CFR 20.1404(a)(4), in that the LTP demonstrates the requirements for public involvement criteria have been met.

- The LTP justifies delaying completion of decommissioning for more than 60 years in accordance with 10 CFR 50.82(a)(3), if applicable.

D. References

1. U.S. Nuclear Regulatory Commission, Regulatory Guide 1.179, "Standard Format and Content of License Termination Plans for Nuclear Power Reactors," January 1999.

2. U.S. Nuclear Regulatory Commission, "Decommissioning of Nuclear Power Reactors" (10 CFR Parts 2, 50, and 51), *Federal Register*, Vol. 61, pp. 39278-39296 (61 FR 39278), July 29, 1996.

3. U.S. Nuclear Regulatory Commission, Regulatory Guide 1.184, "Decommissioning of Nuclear Power Reactors," August 2000.

4. U.S. Nuclear Regulatory Commission, "Radiological Criteria for License Termination" (10 CFR Parts 20, 30, 40, 50, 51, 70, and 72), *Federal Register*, Vol. 62, pp. 39058-39092 (62 FR 39058), July 21, 1997.

5. U.S. Nuclear Regulatory Commission, "Releasing Part of Power Reactor a Facility or Site for Unrestricted Use Before the NRC Approves the License Termination Plan" (10 CFR Parts 2, 20, and 50), *Federal Register*, Vol. 66, pp. 46230-46239 (66 FR 46230), September 4, 2001.

6. U.S. Nuclear Regulatory Commission, Draft NUREG-1757, "Consolidated NMSS Decommissioning Guidance," Vols. 1, and 2, September 2002).

7. U.S. Nuclear Regulatory Commission, Regulatory Guide 1.185, "Standard Format and Content Guide for Post-Shutdown Decommissioning Activities Report," July 2000.

8. U.S. Nuclear Regulatory Commission, NUREG-1713, "Standard Review Plan for Decommissioning Cost Estimates for Nuclear Power Reactors," November 2001.

9. U.S. Nuclear Regulatory Commission, Draft Regulatory Guide-1085, "Standard Format and Content of Decommissioning Cost Estimates for Nuclear Power Reactors," November 2001.

10. U.S. Nuclear Regulatory Commission, Regulatory Guide 1.159, "Assuring the Availability of Funds for Decommissioning Nuclear Reactors" August 1990.

11. U.S. Nuclear Regulatory Commission, NUREG-1307, "Report on Waste Burial Charges," Rev. 9, September 2000.

12. U.S. Nuclear Regulatory Commission, NUREG-0586, "Final Generic Environmental Impact Statement on Decommissioning of Nuclear Facilities," August 1988.

Appendix 1, Acceptance Review Checklist for Unrestricted or Restricted Release of the Site

1. GENERAL INFORMATION

Licensee Name and Address:

Docket Number:

Facility:

- name and address of the facility
- location and address of the site
- brief description of the site and immediate environs
- brief description of any changes to the original site boundary
- summary of the licensed activities that occurred at the site

Description of Site Location and Immediate Environs:

- size of the site in acres or square meters
- State and county in which the site is located
- names and distances to nearby communities, towns and cities
- description of the contours and features of the site
- elevation of the site
- description of property surrounding the site; including the location of all off-site wells used by nearby communities or individuals
- location of the site relative to prominent features such as rivers and lakes
- a map that shows the detailed topography of the site using a contour interval
- the location of the nearest residences and all significant facilities or activities near the site
- description of the facilities (buildings, parking lots, fixed equipment, etc.) at the site and the nature and extent of contamination at the site
- decommissioning objective proposed by the licensee (i.e., restricted or unrestricted use)

2. SITE CHARACTERIZATION (OPERATING HISTORY)

Background Levels Used During Characterization Surveys:

Radionuclides Present at Each Location

- maximum and average radionuclide activities (in dpm/100cm 2, pCi/gm or pCi/l)
- radionuclide ratios, if multiple radionuclides are present

Radiological Contamination Structures, Systems, and Equipment:

- list or description of all structures, systems, and equipment at the facility where licensed activities occurred that contain residual radioactive material in excess of site background levels

- summary of the structures, systems, equipment, and locations at the facility which the licensee or responsible party has concluded have not been impacted by licensed operations, and the rationale for the conclusion

- list or description of each room or area, and equipment within each of the contaminated structures

- summary or map of the locations of contamination in each room or work area

- mode of contamination for each surface (i.e., whether the radioactive material is present only on the surface of the material or if it has penetrated the material)

Characterization Surveys:

- description and justification of the survey measurements for impacted media

- survey results, including tables or charts of the concentrations of residual radioactivity measured

- maps or drawings of the site, area, or building showing areas classified as impacted or not impacted, with justification for considering areas to be not impacted

Surface and Subsurface Soil Contamination:

- list or description of all locations at the facility where surface and subsurface soil contains residual radioactive material in excess of site background levels

- scale drawing or map of the site showing the locations of subsurface soil contamination

Surface Water and Ground Water:

- summary of all surface water bodies and aquifer(s) at the facility that contain residual radioactive material in excess of site background levels

3. & 4. IDENTIFICATION OF REMAINING SITE DISMANTLEMENT ACTIVITIES AND REMEDIATION PLANS

- summary of the radiation protection methods and control procedures that will be employed

- summary of the procedures already authorized under the existing license and those for which approval is being requested in the LTP

- commitment to conduct decommissioning activities in accordance with approved written procedures

- summary of any unique safety or remediation issues associated with remediating contaminated structures, systems, and equipment

- summary of the remediation tasks planned for each room, area and/or system in the order in which they will occur

- description of the remediation techniques that will be employed in each room, area, or system

Soil:

- summary of the removal and remediation tasks planned for surface and subsurface soil at the site in the order, in which they will occur, including which activities will be conducted by licensee staff and which will be performed by a contractor

- description of the techniques that will be employed to remove or remediate surface and subsurface soil at the site

Surface and Ground Water:

- summary of the remediation tasks planned for ground and surface water, in the order in which they will occur, including which activities will be conducted by licensee staff and which will be performed by a contractor

- description the remediation techniques that will be employed to remediate the ground or surface water

Schedules:

- Gantt or PERT chart detailing the proposed remediation tasks in the order in which they will occur

- statement acknowledging that circumstances can change during decommissioning, and, if the licensee determines that the decommissioning cannot be completed as outlined in the schedule, the licensee or responsible party will provide an updated schedule to NRC

5. FINAL RADIATION SURVEY PLAN

- summary table or list of the $DCGL_W$ for each radionuclide and impacted media of concern

- if Class 1 survey units are present, a summary table or list of area factors that will be used to determine the $DCGL_{emc}$ for each radionuclide and media of concern

- if Class 1 survey units are present, the $DCGL_{emc}$ for each radionuclide and medium of concern

- if multiple radionuclides are present, the appropriate DCGL$_W$ for the survey method to be used

- discussion of why the licensee considers the characterization survey to be adequate to demonstrate that it is unlikely that significant quantities of residual radioactivity have gone undetected

- for areas and surfaces that are inaccessible or not readily accessible, a discussion of how they were surveyed or why they did not need to be surveyed

- for sites, areas, or buildings with multiple radionuclides, a discussion justifying the ratios of radionuclides that will be assumed in the final status survey or an indication that no fixed ratio exists and each radionuclide will be measured separately

Remediation Survey:
- description of field screening methods and instrumentation
- demonstration that field screening should be capable of detecting residual radioactivity at the DCGL

Final Status Survey Design:
- brief overview describing the final status survey design
- description and map or drawing of impacted areas of the site, area, or buildings classified by residual radioactivity levels (Class 1, Class 2, or Class 3) and divided into survey units with an explanation of the basis for division into survey units
- description of the background reference areas and materials, if they will be used, and a justification for their selection
- summary of the statistical tests that will be used to evaluate the survey results
- description of scanning instruments, methods, calibration, operational checks, coverage, and sensitivity for each media and radionuclide
- for in situ sample measurements made by field instruments, a description of the instruments, calibration, operational checks, sensitivity, and sampling methods with a demonstration that the instruments and methods have adequate sensitivity
- description of the analytical instruments for measuring samples in the laboratory, including their calibration, sensitivity, and methods with a demonstration that the instruments have adequate sensitivity
- description of how the samples to be analyzed in the laboratory will be collected, controlled, and handled

- description of the final status survey investigation levels and how they were determined

- summary of any significant additional residual radioactivity that was not accounted for during site characterization

- summary of direct measurement results and/or soil concentration levels in units that are comparable to the DCGL, and whether data are used to estimate or update the survey unit

- summary of the direct measurements or sample data used to evaluate the success of remediation and estimate the survey unit variance

Quality Assurance Program to Support Final Surveys:

- description of the QA program management organization, the duties and responsibilities of each unit within the organization, how delegation of responsibilities is managed within the decommissioning program, and how work performance is evaluated

- description of the authority of each unit within the QA program

- organization chart of the QA program

- commitment that activities affecting the quality of site decommissioning will be subject to the applicable controls of the QA program, and activities covered by the QA program are identified in program-defining documents

- description of the self-assessment program to confirm that activities affecting quality comply with the QA program

- commitment that persons performing self-assessment activities will not to have direct responsibilities in the area they assess

Final Status Survey Report:

- overview of the results of the final status survey

- discussion of any changes that were made in the final status survey from what was proposed in the LTP

- description of the method by which the number of samples was determined for each survey unit

- summary of the values used to determine the number of samples and a justification for these values

- survey results for each survey unit including the number of samples taken for the survey unit, and a map or drawing of the survey unit showing the reference system and random start systematic sample locations for Class 1 and 2 survey units and random locations for Class 3 survey units and reference areas

- measured sample concentrations

- statistical evaluation of the measured concentrations

- judgmental and miscellaneous sample data sets, reported separately from those samples collected for performing the statistical evaluation

- discussion of anomalous data, including any areas of elevated direct radiation detected during scanning that exceeded the investigation level or measurement locations in excess of $DCGL_W$

- statement that a given survey unit satisfied the $DCGL_W$ and the elevated measurement comparison if any sample points exceeded the $DCGL_W$

- if survey unit fails, description of any changes in initial survey unit assumptions relative to the extent of residual radioactivity, the investigation conducted to ascertain the reason for the failure and the impact that the failure has on the conclusion that the facility is ready for final radiological surveys; and if a survey unit fails, a discussion of the impact of the failure has on other survey unit information

6.0 COMPLIANCE WITH RADIOLOGICAL CRITERIA FOR LICENSE TERMINATION (DOSE MODELING)

Unrestricted Release Using Screening Criteria:

- For unrestricted release using screening criteria for building surface residual radioactivity, the general conceptual model (for both the source term and the building environment) of the site, and a summary of the screening method

- For unrestricted release using screening criteria for surface soil residual radioactivity, justification on the appropriateness of using the screening approach (for both the source term and the environment) at the site, and a summary of the screening method (i.e., running DandD or using the lookup tables)

Unrestricted Release Criteria Using Site-Specific Information:

- source term information, including nuclides of interest, configuration of the source, areal variability of the source

- description of the exposure scenario used to develop site-specific DCLGs, including a description of the critical group

- description of the conceptual model of the site including the source term, physical features important to modeling the transport pathways, and the critical group

- identification and description of the mathematical model used (e.g., hand calculations, DandD Screen v1.0, RESRAD v 5.81, etc.)

- description of the parameters used in the analysis

- discussion about the effect of uncertainty on the results

- input and output files or printouts, if a computer program was used

ALARA Analysis:

- description of how the licensee or responsible party will achieve a decommissioning goal below the dose limit

- quantitative cost-benefit analysis

- description of how costs were estimated; and, a demonstration that the doses to the average member of the critical group are ALARA

7.0 UPDATE OF SITE-SPECIFIC DECOMMISSIONING COSTS

- cost assumptions used, including a contingency factor and basis for each

- cost estimate addressing the major decommissioning activities and tasks and their relationship to remaining dismantlement activities

- description of the unit cost factors

- estimated costs of decontamination and removal of equipment and structures

- estimated costs of waste disposal, including applicable disposal site surcharges

- estimated transportation costs

- estimated final survey cost

8.0 SUPPLEMENT TO THE ENVIRONMENTAL REPORT

- description of any new information or significant environmental change(s) associated with the site-specific termination activities related to the end use of the site (the environmental evaluation does not have to address decommissioning activities but focuses on site end use)

- description of the impacts associated with those site-specific termination end-use activities, comparing the impact with previously analyzed termination activities, and analyzing the environmental impact of each site-specific activity

- description of proposed termination activities that may result in significant environmental changes that are not bounded by the site-specific decommissioning activities described in the PSDAR, previously issued environmental assessment, or the environmental impact statement

Appendix 2, LTP Areas that Cannot Be Changed Without NRC Approval

The licensee shall implement and maintain in effect all provisions of the approved License Termination Plan [title, version, date] (hereinafter, "LTP"), as approved in the license amendment dated [date], subject to and as amended by the following stipulations:

The licensee may make changes to the LTP without prior approval provided the proposed changes do not meet any of the following criteria:

(A)   Require Commission approval pursuant to 10 CFR 50.59.

(B)   Result in significant environmental impacts not previously been reviewed.

(C)   Detract or negate the reasonable assurance that adequate funds will be available for decommissioning.

(D)   Decrease a survey unit area classification (i.e., impacted to not impacted; Class 1 to Class 2; Class 2 to Class 3; or Class 1 to Class 3) without providing NRC a minium 14 day notification prior to implementing the change in classification.

(E)   Increase the derived concentration guideline levels and related minimum detectable concentrations (for both scan and fixed measurement methods).

(F)   Increase the radioactivity level, relative to the applicable derived concentration guideline level, at which an investigation occurs.

(G)   Change the statistical test applied to one other than the Sign test or Wilcoxon Rank Sum test.

(H)   Increase the Type I decision error.

(I)   Change the documents prescribing the legally-enforceable institutional controls [applies only to license termination under restricted conditions (10 CFR 20.1403)].

(J)   Change the financial assurance instrument(s) required for license termination under restricted conditions [applies only to license termination under restricted conditions (10 CFR 20.1403)].

(K)   Change the alternative criteria approved by the Commission pursuant to 10 CFR 20.1404 [applies only to license termination using alternate criteria].